TROPICAL PLANTS OF SOUTHEAST ASIA

Published by Periplus Editions (HK) Ltd.

Copyright © 1998 Periplus Editions (HK) Ltd.
ALL RIGHTS RESERVED
Printed in the Republic of Singapore
ISBN 962-593-168-6

Publisher: Eric M. Oey
Design: Peter Ivey
Editor: Kim Inglis
Production: Agnes Tan
Additional photographs by Alain Compost (pp 56–57),
Michael Freeman (p 34 top), Jan Kugler ((p 43 bottom), Wayne Lawler (p 26),
Max Lawrence (p 17 top right, p 59), Bill Wassman (p 54)

Distributors
Indonesia
PT Wira Mandala Pustaka
(Java Books-Indonesia)
Jalan Kelapa Gading Kirana,
Blok A14 No. 17,
Jakarta 14240

Singapore and Malaysia
Berkeley Books Pte. Ltd.,
5 Little Road #08-01, Singapore 536983

United States
Charles E. Tuttle Co., Inc.,
RRI Box 231-5, North Clarendon,
VT 05759-9700

Page 4: Amongst the tropical plants here are *Platycerum* (Stag's Horn Fern),
a *Bromeliad* from South America and an *Anthurium*.

PERIPLUS NATURE GUIDES

TROPICAL PLANTS
of Southeast Asia

Text by Elisabeth Chan

Photographs by Luca Invernizzi Tettoni

PERIPLUS

EDITIONS

Introduction

The plants featured in this handbook grow in the region casually called the 'tropics'. Most of them are comfortable around the Equator. A few are more commonly found in latitudes that are described as the 'monsoon tropics' where there is a distinct dry season.

They have been loosely grouped into four categories. The first of these comprises the plants that come to everybody's mind when the words 'tropical plants' are mentioned. They include the ginger and the banana, amongst others.

The next group are those plants that are of great economic significance to man but whose final product, familiar in daily life, does not bear any resemblance to the trees shown in these pages. Included in this class are the rubber and teak trees and the clove.

The third category are plants that are beautiful. Some like the *Amherstia* are rare. Some such as the *Grammatophyllum* are rarely seen in bloom. Others are beautiful even when seen in their species form which may be obscured by hybridization.

There are oddities in every sphere of life and the plant world is no different. Some such as *Rafflesia* are singular. It is a peculiar flower, restricted in its distribution, and hardly ever seen. The *Nepenthes*, more often seen in cultivated form, has pitchers in its many species that are far more dazzling.

Before the present era of cheap, mass travel, the tropics had a mystique that no longer exists. Plants from these places were familiar only from illustrations in travel or botany books. Some of these old line drawings and lithographs which antedated the camera are reproduced in the pages of this book.

Many of the plants in this book are world travelers. In the centuries beginning from the 15th when European exploratory voyages began in earnest, plants of economic and botanical interest were taken from one hemisphere to the other. Even before this, many significant plants had been taken from their point of origin to other places. Many of the plants now common to tropical Asia are exotics in the true botanical sense. The story of rubber is a classic example of a valuable plant that was developed far from its native home.

There are now international agreements governing the transportation of plants because the plants themselves may be endangered and because pests and diseases may be carried far beyond their native homes into areas where there is no balance with their natural predators.

Sometimes travelers who are not eco-tourists to the tropics may be disappointed if they wish to see or read about tropical plants. They are usually short of time and have to visit a Botanic Garden, a Nature Reserve or a plant market or nursery. All these visits may not be feasible. We hope that this modest guide may make up for some of these disappointments.

Screwpine

Pandanus spp.

Botanical family:
Pandanaceae

Pandanus tectorius (see left and far right) is a very common sight along Old World tropical beaches. Formerly known as *P. odoratissimus*, it is a medium tree-like plant. It has a trunk and very obvious prop roots and its leaves are gathered in a large, stiff clump at the top. Often there is a fruit looking somewhat like a pineapple. It is a multiple fruit, an attractive orange or red when ripe, made up of clearly defined carpels which when fully ripe, break off, and—if not eaten by an animal—can be seen lying on the beach waiting for the tide to take them to another spot to germinate. *P. tectorius* is often planted in seaside gardens because of its ability to withstand strong, salty, sea winds.

The leaves are long, dark green and very stiff, with spines along the edges and, in some species, along the midrib. They are spirally arranged on the trunk, hence the common name. These leaves are of economic importance in village and cottage industries, where the work is done mainly by women. Initially the leaves are cut and the spines stripped off (a painful procedure). Then the leaves are torn into various widths, dried and tied in hanks, dyed with synthetic dyes, and woven into mats, baskets, hats, trays and other objects for sale in tourist markets.

Pandanus amaryllifolius (opposite middle left), is a domesticated species without spines of which many forms have been cultivated, including one called the *Pandan seran*i the 'Eurasian pandan' in the Malay language. The origin of this name is obscure. *P. amaryllifolius* is a small species of about 50 cm characterized by very sweet-smelling leaves. It is cultivated extensively throughout India, Sri Lanka and Southeast Asia to provide flavoring for food. It is also grated very finely into a pot-pourri mixture for temple and altar offerings and, rather peculiarly, to be placed in drawers and taxis to deter cockroaches.

The species on opposite top left is *P. spiralis* which has a marked twist to its habit of growth, and the variegated form, opposite bottom, is a cultivated ornamental.

Tree Fern

Cyathea spp.

Botanical family:
Cyatheaceae

Tree ferns are forest dwellers, partial to dim coolness and damp. Some are found in lowland forests where they do not grow very tall (up to 3 m). The tree ferns found in the hills and upland areas of the tropics do very much better and are a conspicuous feature of these areas. They also grow in more open secondary forest.

The tree fern resembles a delicate palm tree. It has a rough and hairy trunk, marked with the remains of leaf bases. The fern fronds at the growing tip look like palm fronds as they are concentrated at the top of the trunk.

Within the genus they are easy to pick out, but to identify species requires a deeper knowledge of the scaly young parts and leaves. A clue may come from the name, taken from the Greek word for cup, *kyatheion*: this is what to look for in the cover of the spore cases that are found on the lower side of the fern frond.

Tree ferns are often logged for the horticultural trade as the trunks make good growing poles for epiphytic plants.

Bird's Nest Fern

Asplenium nidus

This is a popular house plant in temperate climate homes where it forms a dainty roseate of about 10 cm. In its native Old World tropics it can grow to 2 m across.

Botanical family: Aspleniaceae

In its natural habitat, *Asplenium nidus* varies in size depending on its location and the availability of water and a continuous supply of decomposing detritus from trees and insects which provide the fern with nutrients. This matter falls into the 'nest' formed by the fronds which grow in alternately overlapping circles.

A colonizing fern, it is epiphytic on suitable trees, but is commonly seen on the ground where it has fallen from its perch but continues to grow. As the fern grows upward and outward simultaneously, the root mass becomes deeper and spongier and is able to hold a great quantity of water. This attracts other fern spores to colonize the root mass. In some Malay rural areas, the plant is believed to have supernatural properties, or of being the home of the *lansuir*, a female banshee hostile to pregnant women.

Botanical families:
Costaceae;
Zingiberaceae

Gingers

*Alpinia purpurata; Costus speciosus; Etlingera
elatior; Tapeinochilus ananassae; Zingiber officinalis*

Gingers are among the flashier members of the plant
world. They have a high profile in coffee-table books and
are the mainstay of the tropical cut flower industry.

Costus speciosus (see opposite centre left) is a member
of the family Costaceae. The fleshy, velvety leaves grow
spirally around the stem in a characteristic clockwise
direction. The stem is cane-like, reddish and tall, and the
dark red inflorescence is terminal on the stem. The white
or slightly pink flower, tubular with a pretty frill, emerges
from the bracts and is edible. The rhizome is used for
medicinal purposes.

Tapeinochilus ananassae, also from the family
Costaceae, is appropriately known as the wax ginger (see
far right). The plant produces canes that are dark red in
color and tall, but, unlike the *Costus speciosus*, it branches
at the top of the stem. The inflorescence has large
bracts—red, shiny, stiff and pineapple-like—with insignificant flowers. It does not grow tall; several cluster around
the base of the plant.

Alpinia purpurata (on bottom right) is a *zingiber*. It is a
popular landscape ginger, the pink form of which is called
'Eileen Mcdonald'. The plant propagates itself by produc-
ing plantlets from the flowering bracts.

Etlingera elatior (on opposite top left) belongs to the
Zingeberaceae. Called the torch ginger, its leafstalks are tall,
about 10 to 15 m and the 'torch' itself is about 1 to 2 m.
Torches, produced singly from the ground on a long stalk,
are composed of big flower heads of waxy overlapping
bracts with small flowers. In Southeast Asia they are vis-
ited by small sunbirds, as each row of flowers opens in
turn. The bud is used sparingly as a flavoring for food.

Zingiber officinalis, the oldest 'Oriental' spice known to
the Western world is the familiar ginger rhizome used
worldwide (see above left). Although it was recorded in
China 500 years ago, its place of origin is unknown, but it
is suspected to be an Indian cultigen.

Banana Plant

Musa spp.

Botanical family:
Musaceae

Any discussion of tropical plants must include the banana. Yet ethnobotanists do not know exactly where the fruit originated. The most generally accepted theory is that as the Indo-Malesian area is the main center of diversity, this is strong evidence of origin.

There are hundreds of edible banana varieties; in Indonesia alone there are over 230 recorded, but the bananas of commerce are far fewer, due to considerations of quality, aesthetic appeal, flavor and so on. The two species banana that are considered to be the parents of most of the edible seedless bananas eaten by man are *Musa acuminata* and *Musa balbisiana*.

The banana is such a pan-tropical that it grows everywhere man has planted it. It is even said that the slaves in the West Indies for whom the breadfruit was painfully collected by Captain Bligh, declined to eat it because they preferred bananas.

Species bananas are more interesting botanically, even though they are tasteless or full of seeds. Pictured here are three of the more spectacular varieties. The banana with the incredibly long fruiting stem is called the 1000 banana plant in Indonesia and Malaysia. Even if the actual number of bananas does not reach 1000, the bunch is a marvel of nature: It is sometimes so long and the fruits so numerous that the bunch reaches the ground.

The banana pictured on the top left is *Musa velutina*. The pretty red fruits actually peel themselves from the base of the fruit up, to entice a potential customer to eat them and thereby spread the seeds. The banana (opposite middle left) with the striking red flowers, *Musa coccinea*, is purely ornamental as its fruits are small and hard.

In addition to being eaten fresh, bananas may be cooked, chipped, made into alcoholic drinks or processed into starch. The leaves are used to wrap foods or to line utensils in which food is prepared. The flowers of the inflorescence and the center of the stem are also edible.

12

Cordyline

Cordyline terminalis

Botanical family:
Dracaenaceae

Cultivated for their beautiful leaves, cordylines are native to Australasia, the Pacific Islands and tropical America. The center for cordyline development is Hawaii where it is known as the *ti* plant. In Asia, collectors look to Thailand.

Most cultivated cordylines have long leaves that vary in width and length. They grow in a roseate form at the top of a stem that is marked all the way up the stem by the scars of the leaf sheaths. Some newer hybrids have leaves twisted like corkscrews, and many other cultivars have broad leaves or leaves that are almost cup-shaped. In all cordylines, the flowers are inconspicuous. The leaves come in all colors except blue, black, pure yellow, pink or other delicate colors. There are many variations on the rose-purple color, and there is a striped green and white form.

The landscaping style created by Roberto Burle Marx, now popular in public landscaping all over the tropics, has made the cordyline an essential element. Planted in blocks of a single color, they are striking and easy to maintain.

Codiaeum

Codiaeum variegatum

Codiaeum are found in almost all tropical gardens, parks, roadside plantings, Botanic Gardens and cemeteries. They are popularly and inaccurately called 'crotons'. They are almost trouble free and are very colorful ornamentals. The flowers are inconspicuous, but the shiny leaves come in innumerable variations of color, shape, size and form.

Botanical family:
Euphorbiaceae

The forms commonly seen are cultivated varieties of plainer, duller, single color species now rarely seen. As they are so variable, the only way to ensure that a plant is the desired one is to propagate it vegetatively.

Codiaeum are indigenous to Malesia and the Pacific. They come in all colors except blue, black, pink and purple. There are many named varieties, and recently, Thai horticulturalists have developed dwarf varieties suitable for the pot plant culture of urban areas. Codiaeum are useful as specimen plants, hedges and as part of a general planting. The taller specimens reach about 2 m and they require full sunlight. They can give a garden a 'hot' tropical look.

15

Botanical family:
Araceae

Aroids

Alocasia macrorrhizos; Amorphophallus paeoniifolius;Caladium x. hortulanum; Monstera deliciosa; Pistia stratiotes

Araceae are a large botanical family classified at present into 2,500 species. The plants pictured here belong to a tiny part of them. They do not resemble each other in any way noticeable to a casual observer, but they illustrate the vast range of tropical aroids that exist.

All aroid inflorescences have a spathe which is the more noticeable part and a spadix which is the upright organ that carries the real, but insignificant, flowers. Some are absolutely weird and wonderful, like the *Amorphophallus paeoniifolius* (opposite top right). Its inflorescence, when it appears, is the subject of a news report not only for its size and rarity, but also because of its overpowering smell; others, like the *Athurium* (see left below), the staple of the cut flower trade, are moderately flashy; some are totally undistinguished like that of the *Caladium* shown opposite bottom; and some like the *Monstera deliciosa* have a large inflorescence that produces an edible fruit which tastes like a combination of all tropical fruits together. The pretty roseate *Pistia stratiotes* (opposite middle left) has hidden its inflorescence from anyone without a hand lens.

Aroids also have stunning leaves: The *Monstera deliciosa* is called the Swiss cheese plant because it has holes in its leaves. The plant shown above left is a variegated form. Caladiums have beautifully marked leaves: *Caladium x. hortulanum* (opposite bottom) is thought to be a parent of many of the hybrids now available. It was introduced in the Old World tropics in the last century as an ornamental and quickly escaped into the wild. The leaf stems of the *Amorphophallus* group are marked like pythons.

Aroids have many uses and are found in many different habitats. *Pistia stratiotes* is a specimen that grows in water. It oxygenates the water, but can become a pest of waterways because it spreads fast. The *Alocasia macrorrhizos* on opposite top left is a common species seen in wasteland as well as in gardens.

Tongkat Ali

Eurycoma longifolia

Botanical family:
Simarubaceae

The Malay name for this plant translates as Ali's Walking Stick. It is a small tree or leggy shrub, up to 5 m in height, often without branches. It has a sparse appearance, with coarse, dark bark, a feathery leaf cluster at the top, and a well-developed root system.

Its many small, scented flowers appear in long, pendant panicles, followed by attractive, bright red fruits about 1.5 cm long. The plant can be propagated by seed.

Tongkat Ali is found in wasteland, poor, dry or sandy soils, and also as an understorey plant in secondary forest. It used to be very common in Indochina and in the Malay Archipelago and was traditionally used in Malay medicine for fevers and as a tonic.

It has recently become the target of reckless and destructive collection because of untoward publicity about the reputed aphrodisiac properties of its bark and roots. This is an old belief that has never been scientifically substantiated, but that has resurfaced to the plant's detriment.

Datura

Datura sp.; Brugmansia sp.

Botanical family:
Solanaceae

Datura is an herbaceous plant with erect flowers, while *Brugmansia* is a woodier shrub in the same family with pendulous flowers. Their habit of growth is picturesquely called candelabriform. The flowers are strikingly beautiful, but all parts of the plant are intensely poisonous especially the seeds. The plants have been hybridized into horticulturally superior forms from the original wild, weedy ones.

The flowers bloom freely and are long and tubular, flaring out into points. Strongly fragrant, they are pollinated by a night-flying moth. *Datura* flowers are white, purple or yellow, whilst those of *Brugmansia* are white, peach and, in one species, red and yellow. The fruit is an ovoid capsule, either spiny or smooth. The seeds resemble tomato and capsicum seeds, indicating the same botanical family.

Found in the wild as well as in cultivation, Datura was associated with South American Indian rituals, and in India and Southeast Asia, it has been documented in criminal poisoning activities.

Anatto

Bixa orellana

Botanical family:
Bixaceae

This small tree or large shrub originates from tropical America, but it is so decorative and useful that it has spread around the tropical world. It has large, thin, heart-shaped pointed leaves and five attractive white or pink flowers in a panicle. It flowers profusely and frequently in all its adoptive homes.

The dark red or crimson fruits that follow the flowers are even more eye-catching. They appear in large bunches, each pointed capsule of roughly ovoid shape about 5 cm and covered with short, stiff spines. They are not edible, but the seeds in them are of commercial value.

When the fruits are dry, they turn brown and split open revealing the seeds inside. The seeds have a bright red covering, which when extracted, provides an orange or yellow dye. The dye is used in Filipino and Thai cuisines as a food coloring and is also used extensively commercially to color dairy products, margarine, cosmetics and confectionery. In Sri Lanka it is mixed with lime juice to dye cotton mauve.

Blue Pea

Clitorea ternatea

This very pretty blue, azure or white flower of obscure origin is popular as a garden climber. It is a slender plant with light green leaves, belonging to the pea or bean family. Its seed pod shows this relationship very clearly. The flowers are not very large, about 3 or 4 cm long, but they are presented face up with the keel uppermost and the standard below. Linnaeus thought it resembled the female genitalia and thus gave it its generic name. In Victorian India, it was more delicately referred to as the mussel shell flower. Its specific name *ternatea* is a misnomer, as it does not originate from the Indonesian island of Ternate.

There are single, semi-double and double forms, the double flowers being the prettiest. The plants seed freely around the mother plant, and it is seen at its best on a trellis, planted in a mass so as to make an impact.

The deep blue flowers are also dried and used as a dye for food coloring throughout Southeast Asia. This same blue dye is also used in Indonesia to dye textiles.

Botanical family:
Leguminosae

Coconut Palm

Cocos nucifera

Botanical family:
Palmae

This is the quintessential tropical tree, yet no one is certain where it originated. Its wide dispersion is a result of the nut floating safely in its waterproof husk in sea water, and the fact that man, in his global voyages, took it with him.

The tree may have derived its name from the Portuguese word *coco* meaning bugbear or ugly mask. This came from the fact that the husked nut is supposed to resemble the face of a monkey. The husked nut has three marks on it, which correspond to the eyes and mouth of the monkey and the only place that one can pierce the coconut is through the mouth. This notion is still firmly believed in Sri Lanka.

Every part of the coconut is useful, including the roots which are used in medicine. Coconut water straight from the nut is sterile and can be used to clean wounds. The white 'milk', squeezed out of finely grated fresh coconut meat, is full of rich oil. In addition, an enormous number of utilitarian articles are made from this tree.

Traveler's Palm

Ravenala madagascariensis

In a highly stylized form, the traveler's palm is used as the logo for the Raffles Hotel in Singapore. The characteristic fan-shaped silhouette has an enduring connotation with the 'exotic East'. However, as its name indicates, it comes from Madagascar. Another misnomer is that it is not a palm, but a member of the banana family.

Botanical family:
Strelitziaceae

R. madagascariensis is associated with travelers because the bases of the leaf sheaths contain water which may be drunk by thirsty travelers with no other source of water. This may be apocryphal as it is a tall plant, reaching about 15 m under optimum growing conditions.

The tree has a large inflorescence that strongly resembles the popular cut flower, the *Strelitzia reginae*, of the same family. The inflorescence is not produced freely in the very wet tropics but is common in places such as Thailand. In a domestic garden, the *R. madagascariensis* is often planted at an angle so that its large fan will shield the house from the hot afternoon sun.

Coco-de-Mer

Lodoicea maldivica

Botanical family:
Palmae

The *Lodoicea* palm is indigenous only to the slopes away from the coast on the islands of Praslin and Curiense in the Seychelles. It was named by Rumphius after the daughter of King Priam. Before it was discovered in 1768 it was known only from fruit washed up in the Maldives, hence its specific name *maldivica*. As sea water kills the seed, myths grew up around it, among them the legend that the tree grew in the sea, hence the common name.

It is a solitary and very slow-growing palm, reaching its full size of 25 to 30 m after 200 years. The leaves are immense, fan-shaped, leathery and a grayish green. The female tree does not flower and bear fruit until it is 25 to 30 years old. The fruit containing a bi-lobed seed is the largest seed in the world and takes five to seven years to mature and a further six to seven years to germinate. Mature 'nuts' weigh about 20 kg each.

There is a soft, edible flesh within and the shells are used as containers and ornamental boxes (see page 45).

24

Sugar Palm

Arenga pinnata

Arenga pinnata is the sugar palm of Malaysia and Java that is said to give a rich sugar. It is a large, solitary palm that can attain a height between 12 and 20 m. Indigenous to Southeast Asia, it grows wild, but is sometimes cultivated. The leaves are gray-green, with a whitish cast beneath. The base of each leafstalk is covered with a tough and weather-resistant, black fiber called *gemuti*, which is used in thatching and the manufacture of brooms and rope.

This palm flowers from the topmost leaf sheaths downwards. When an *arenga* is seen with the flower stalk near the base of the tree it is a sorry sight, as it is clear that it is dying. It lives for about 15 to 20 years.

The golf ball sized fruits are produced in an enormous bunch. The unripe fruit contains an edible endosperm which is extracted and boiled to make a sweetmeat called *kabung*. The sugar is obtained by bruising the inflorescence and collecting the sap which oozes out. The juice is reduced to sugar by boiling and poured into molds to set.

Botanical family:
Palmae

Nypa Palm

Nypa fruticans

Botanical family:
Palmae

Nypa fruticans is a monotypic species, which means that it is the only species of its genus in the world. It is also possible that it is the oldest palm in the world, as fossils dating from 100 million years ago have been found in the Northern Hemisphere. Its present distribution is the Southeast Asian tropics, New Guinea, Australia, Bengal and Sri Lanka. It is found in swamps along the banks of tidal rivers in brackish water.

Nypa has long upright leaves up to 9 m long, each forming a crown and it develops into a clump by its creeping rhizomes. The orange inflorescence has both the male and female flowers together; the female flowers are held at the top of the column and the male flowers are on separate stalks around them.

In Southeast Asia *Nypa* palm products have been used for hundreds of years. Old palm fronds called *attap* are the preferred material for roofing, and it may be made into baskets, brooms and many other utilitarian items.

Sago Palm

Metroxylon sagu; Metroxylon rumphii

Sago, or *sagu* in the Malay/Indonesian language, is one of the principal substitutes for rice in those parts of tropical Southeast Asia which have insufficient rain for wet rice. The palm tree from which sago starch is obtained comes in two forms, spiny (*M. sagu*) and smooth (*M. rumphii*). It forms clumps both in cultivation and in the wild. The natural habitat is lowland freshwater swamp and they are found throughout the Asean countries to New Guinea.

The palm builds up its stores of starch over its life of about 15 years and attains its maximum store just before the inflorescence opens. The palm is felled, the trunk cut into many lengths and the pith is manually extracted and processed. The purified starch is then dried and preserved, notably as flour and baked biscuit. It is also used in the textile and pharmaceutical industries.

Botanical family:
Palmae

The pearl sago of commerce is this same starch mixed again into a paste and sieved through mesh of various sizes. The finished sago pearls have a long shelf life.

Lontar Palm

Borassus flabellifer

Botanical family:
Palmae

This tall stately palm, called the palmyra or tal west of the Malay Archipelago and the lontar in Indonesia, ranges from tropical Africa through India and Sri Lanka, Burma, Southeast Asia to east Indonesia.

It is a solitary, erect palm with a solid trunk covered when young with the remains of leaf sheaths. The leaves are stiff and fan-shaped and the outline of the crown is spherical. The fruits (nuts) are black or dark brown and are produced in a long bunch. The palm prefers a climate with a distinct dry spell.

A Tamil song allegedly records 801 uses for the lontar. Two articles made from the leaves, namely a bucket and a musical instrument like a mandolin, are not made from any other tropical palm. The palm leaves have been used for writing since 450 AD. To form a manuscript, the leaves are cut into rectangular strips, then threaded together. The fruits can be made into a drink, the soft kernel ('sea coconut') is eaten, and the germinated nuts are also edible.

Talipot Palm

Corypha umbraculifera

This palm is a distinguished one. Not only is it the tallest of all palms, it has the largest inflorescence of any plant in the world and it is the national floral emblem of Sri Lanka. The name *talipot* is derived from two Hindi words which mean the leaf of the tall tree. It is suspected to be a cultigen. Native to Sri Lanka, it is cultivated there as well as in South India. It is never found in the wild and is a specimen palm in tropical Botanic Gardens worldwide.

The palm reaches maturity after 30–50 years. It then produces an inflorescence which is a continuation of the stem, a staggering 6 m in height and made up of many branches. There are over a million small flowers and eventually about 250,000 fruits. As the fruits ripen and fall, the palm, exhausted by this tremendous effort, dies.

In its native Sri Lanka as well as in India, the leaves have been used in the traditional way to record personal horoscopes and religious books. Other domestic articles, such as umbrellas and thatch, are also made from the leaves.

Botanical family:
Palmae

Jackfruit

Artocarpus heterophyllus

Botanical family:
Moraceae

This fruit has the distinction of being the biggest in the world. It is a compound fruit, many little fruitlets coming together to form a large whole. The tree is quick growing and is found in villages, fruit orchards and suburban gardens where it begins to fruit from about 18 months to 2 years.

It is a popular fruit with orange-yellow flesh. The seeds, roasted and boiled, are slightly sweet and resemble the seeds of their near relative, the breadfruit.

The tree which can reach 15 m when fully grown is kept pruned to a more manageable height in cultivation. The fruits appear from the trunk and the main branches and are carefully wrapped in paper or jute bags as they are a great attraction for fruit flies. The leaves are used in agricultural societies as goat feed, and the fine yellow-colored wood is used for furniture and cabinet work, particularly in Sri Lanka. The greatest drawback to the fruit is that it produces copious amounts of a sticky white latex that is persistent and may cause allergic reactions.

Breadfruit

Artocarpus altilis

The breadfruit tree is a handsome addition to a large garden, roadside or park. It is tall, up to 20 m, with very large, dark green, shiny leaves that turn tan-colored before falling. They are leathery and deeply indented, holding their shape well for flower arrangements. It is native to the South Pacific, but has become common in all wet tropical countries. It was the staple diet of the Polynesians.

Botanical family:
Moraceae

The seedless variety shown below is a cultivar that has been vegetatively propagated for food and is not found in the wild. The specimen (above right), sometimes called the breadnut, is a wild seeded form with edible seeds.

Generally, breadfruits are round, about 30 cm, and bright green with gummy streaks on the skin surface. They are borne at the ends of branches in twos or threes. The pulp is quite palatable. Historically, breadfruit was collected by Captain Bligh (of the *Bounty* fame) as it was deemed appropriate food for slaves on slaveships. Today, Caribbean cookbooks still carry breadfruit recipes.

Keluak

Pangium edule

Botanical family:
Flacourtiaceae

In many Southeast Asian markets one can see in jute sacks, large brown seeds with a grayish cast, about 5 to 6 cm across in a flattened, ovoid shape. They are called *buah keluak* in the Malay and Indonesian language and are a delicacy.

These are the seeds of one of the world's wholly poisonous trees. Indigenous to the Malay Archipelago, *Pangium edule* is a medium to tall tree with sizable, long-stalked leaves, arranged in a pyramidal manner, enabling all leaves to benefit from the sun. It is cultivated in villages and is semi-naturalized.

The seeds are contained in a fruit that has a distinct nipple at the tip. When the fruit is ripe, it is left until the pulp falls away. The seeds are then taken out, washed and boiled. They are buried in ashes in a pit covered with earth and banana leaves and left for 40 days, after which they are dug out and cleaned—ready for market.

The seeds are cooked in a spicy curry usually with chicken by Malays, Indonesians, Eurasians and Peranakans.

Durian

Durio zibethinus

Botanical family:
Bombacaceae

The king of tropical fruits has been maligned ever since explorers to the Malay Archipelago put about the rumor that *Durio zibethinus*, while tasting of all sorts of delicious flavors, had a repugnant smell. It actually has a fresh, although admittedly metallic, smell.

The durian belongs to quite a large botanical family, but it is the most famous member. There are many species, but *D. zibethinus* is the species that has been hybridized so that it can be grown easily in a garden.

Such a famous fruit has many fables attached to it. The most popular one is that it is an aphrodisiac. It is said that the durian must be eaten with the queen of fruits, the mangosteen, and if after a durian feast you do not wish anyone to know you have eaten it, you must drink water out of the fruit skin which has deep curved sections.

Jungle animals love the fruit. Elephants roll the durian delicately in dry leaves to insulate the thorny skin then gently press it open with a foot.

Opium Poppy

Papaver somniferum

Botanical family:
Papaveraceae

The opium poppy is cultivated legally and illegally in a wide band of the semi-tropical to monsoonal tropical regions of the world from Iran to China.

Opium, heroin and morphine are the three useful medicinal products of *Papaver somniferum* that can alleviate misery and pain, but paradoxically may cause misery and pain through abuse. These three are the most familiar of the 25 or so different alkaloids contained in the plant. Opium has caused political upheavals, and its derivatives—morphine and diamorphine—are two substances which concern narcotics and crime fighting agencies worldwide.

The plant is a subshrub that produces a brilliant red flower. Opium is the dried latex from the immature, green, seed capsule. It is obtained by carefully scoring each seed capsule longitudinally so that the sap oozes out and dries on contact with the air.

The seeds of the poppy are used in baking as decorative embellishments and for flavoring in curries.

Clove Tree

Syzygium aromaticum syn.
Eugenia caryophyllus

Cloves were known to the Chinese in the 3rd century BC and to the Egyptians since the 2nd century AD. From the 15th century, European powers fought bloody wars to control the trade in cloves and other spices. This resulted in permanent political and cultural changes to the Spice Islands which were at the center of the clove trade.

Cloves trees are indigenous to only five tiny volcanic islands in the Moluccas islands in Indonesia. They are medium sized, with scented leaves with a wavy margin. They have a pinkish tinge when young. The familiar brown clove is the young bud produced in clusters that is harvested and dried when the greenish-white buds develop a pink flush. The tree begins to flower when about five years old and reaches full bearing age at 20 years.

Cloves are used worldwide for flavoring and preserving food, forming a small but crucial part of many dishes. The essential oil, eugenol, has medicinal applications. In its native Indonesia, it is the flavoring for the popular *kretek* cigarette.

Botanical family:
Myrtaceae

Kapok Tree

Ceiba pentandra

Botanical family:
Bombacaceae

This extraordinary looking tree (pronounced kar-pohk) has been described as a tree designed by a Public Works Committee. Its trunk rises straight up until about 30 m, then tapers at the top, with tiers of branches held stiffly horizontal. The pale gray trunk may or may not bear short spines. It is not particularly leafy.

The flowers are bat-pollinated. The fruits are pods, hanging down vertically from the branches, giving the tree a rather sinister appearance. It is easily propagated by sticking branches into the ground and is commonly used for boundary marking and casual fencing.

Kapok is a creamy-white, silky floss up to 3 cm long surrounding small, black seeds. When the pods are ripe, they burst open dispersing the floss. The floss is used as a stuffing for bedding and for insulation purposes. In Malay folklore, the tree is inhabited by a female vampire, the *pontianak*, who preys on men and babies. A nail, hammered into the trunk, is supposed to prevent her from roaming.

Teak Tree

Tectona grandis

One of the old images of Burma and Thailand is that of a trained elephant hauling teak logs deep in the tropical forest. There is now no more logging and teak is grown in managed plantations in other Asian countries as well.

Tectona grandis is the species from which commercial timber is obtained. It is indigenous from India to Laos, and was introduced 400–600 years ago into Indonesia, where it is now naturalized.

Botanical family:
Verbenaceae

The tree is tall, up to 50 m, and the leaves are the size of a dinner plate. The flowers are inconspicuous and the seeds are surprisingly small. The sap wood is white and the heartwood is golden and aromatic.

This is the most popular wood in Asia for furniture, joinery, house-building, ship-building, bridges and veneers. It is an all-purpose utility timber, a semi-hardwood, that is easy to work. The tree has a straight, even and well-formed bole. It is believed to resist attack from termites and fungi because of its oil content.

Botanical family:
Palmae

Oil Palm

Elaeis guineensis

The oil palm produces palm oil and palm kernel oil. It is a native of tropical Africa that is extensively planted commercially in many other tropical countries. The oil from this palm is the mainstay of the margarine and soap industries, and has many other applications.

Palm oil is produced from the pulp of the fruit and palm kernel oil from the seed. The oil from the fruit is light yellow to red in color and is rich in carotenes, the precursor of vitamin A. Palm kernel oil is almost colorless and, in addition to margarine, it is used in confectionery, ice-cream and in the baking industry. The residue left after processing is used for animal feed and the shells of the kernels are used for horticultural purposes.

Oil palm trees have a heavy look among palms. The trunks are ringed with scars of leaf sheaths and the palm leaves are a dense crown of dark green, protected by thorns. The fruits in big, orange-red bunches are attractive and have an oily feel.

Rubber

Hevea brasiliensis

Botanical family:
Euphorbiaceae

The two Englishmen associated with the rubber industry based on this Brazilian tree are Henry Wickham and Henry Ridley. Allegedly in defiance of the Brazilian ban on the export of seeds, Wickam obtained the seeds of the *Para* rubber tree as it was then known. However, he may have had clearance from the local Customs to take out 70,000 seeds. The seeds were germinated at Kew during the late 19th century, and the seedlings eventually sent to the Singapore Botanic Gardens. Ridley, the then Director, tirelessly encouraged planters in the Malay States to plant the new discovery, earning him the soubriquet 'Mad Ridley'.

The result has been the vast rubber estates of Southeast Asia. The tree that produces the white latex that is rubber has come a long way from the seedlings grown by Ridley. There have been many improvements in the stock and research continues. To extract the rubber, a 'Y' incision is carefully cut in the bark of the tree at dawn, and the latex drips are collected in cups and processed.

Bamboo

Bambusa vulgaris; Dendrocalamus asper; Gigantochloa atroviolacea

Botanical family:
Gramineae

In 1903, a researcher counted the number of uses to which bamboo was put, and the total amounted to 1,546. Even if some of these are obsolete, there are still a tremendous number of uses for these plants belonging to the grass family.

Bamboo has formed an integral part of the lives of many people for several millenia. It was recorded in India in 1000 BC. The most important use for bamboo is as a building material, but bamboo as food and as food preparation utensils also has a long history. It has been celebrated in paintings, poetry, handicrafts, musical instruments and in aphorisms in many cultures. Bamboo is now managed as a renewable resource in plantations.

Above:
Dendrocalamus asper;
Bottom left:
Gigantochloa atroviolacea;
Bottom right:
Bambusa vulgaris

As for species, there are estimated to be about 1,000 of which Southeast Asia accounts for 200. The fastest growing bamboo in the world originates here. There are green, yellow and stunning black forms. A bamboo grove is a cool peaceful place with a faint shushing noise.

Rattan

Calamus sp.; Daemonorops sp.

Botanical family:
Palmae

Rattans are mostly climbing members of the palm family —lianoids—although there are short-stemmed or clump forms. They have been used in the culture of the Asian tropics for thousands of years. There are many species and every one of these is thought to have been used by man.

Armed with vicious thorns, on their stems and as extensions of their leaves, rattans claw their way up the surrounding arboreal forest towards the sun. Unlike the bamboo, their stems are solid, and in the wild they can reach many meters in length. Rattan has strength; it is durable, pliable, cheap and the stems are a uniform size. A young shoot will be of the same diameter when mature.

Up to recent decades, all rattan was collected from the wild. Today there are managed plantations. The uses of rattan are many: Some now have an archaic flavor—for example the construction of sedan chairs. Another more controversial use is for corporal punishment with the rattan cane for criminal offenses in certain jurisdictions.

Cocoa

Theobroma cacao

Botanical family:
Sterculiaceae

The cocoa tree is from South America. It was first discovered in Mexico, where the Spaniards found the Aztecs drinking *choclatl* from where the word 'chocolate' is derived. It translated as 'Drink of the Gods', a translation that Linneaus rendered into Latin as *Theobroma*.

The products of this tree—chocolate, cocoa and cocoa butter—are a vital part of the economies of countries where it is presently grown. These are, among others, Ghana, Malaysia, Sri Lanka and the West Indies; the cocoa plant is now a pan-tropical.

Theobroma cacao is not a large plant. It has large, limp-looking leaves and produces on its solid, woody trunk and branches, hundreds of small flowers. From this extravagance of bloom relatively few pods develop. They are red or yellow and are hard and ridged, about 20 cm long. The seeds inside, the beans, are the violet-colored cocoa beans. These are processed to produce various medicines, skin care products, chocolate and other foodstuffs.

Coffee

Coffea spp.

Coffee originated in either Kenya or Ethiopia. The Arabs introduced *qahwah* to the world and the Dutch and French spread its cultivation to their respective colonies.

Botanical family:
Rubiaceae

The three commonly grown species of coffee are C. *arabica*, reputedly the best, C. *canephora* (syn. *robusta*) and C. *liberica*. Instant coffee is a blended product.

Coffee bushes make decorative garden plants, though some pruning is required to keep them shapely. They have glossy, dark green leaves, and their flowers are starry and white, with a very sweet fragrance that carries on the wind. The berries are bright red or rust, depending on the species. The coffee 'bean' is the seed within the berry. There are two beans per berry, and each has a characteristic groove down one side. A healthy plant produces enough berries to provide a home with its own coffee. Home gardeners estimate that it takes about 20 beans per cup.

The appeal of coffee as a stimulant is derived from alakaloids, principally caffeine and theobromine.

Botanical family:
Palmae

Betel Nut Palm

Areca catechu

The betel nut palm has given its Malay name, *pinang*, to the island of Penang.

A truly domesticated, cultivated plant, it is a slender, graceful, solitary palm with a dark green, compact crown. It reaches fruit-bearing age at about six years and continues to fruit for 25 more. At maturity, it is 18 m and is a pretty sight, with its fruits held in bright orange-yellow clusters.

The betel nut inside the fruit is the famous masticatory chewed throughout the tropical East. The fruit is dried, the husk removed and the hard nut inside is pared into thin slices, mixed with lime, tobacco and, depending on taste, gambier and cloves. The whole is then wrapped in the leaf of *Piper betle* ready for a 'chew'. The effect is stimulating and an appetite suppressant. It also produces a characteristic red staining in the mouth.

The roots, leaves and fruits of the palm are used in traditional Malay medicine. The powdered seed has also been used as a worm mixture for dogs.

Betel Leaf Vine

Piper betle

The betel vine, *Piper betle*, is seen wherever the *Areca* nut palm is cultivated for the betel nut chewing habit. They are used together for a mild stimulating effect. The vine is a neat, pretty one with lighter green, heart-shaped leaves than its relative the black spice pepper, *P. nigrum*. The leaves are used as a wrapping for the 'chew' or 'quid' based on the areca nut. They have a sharp, 'hot' taste and are reputed to have sustaining properties.

Botanical family:
Piperaceae

The mature leaves are picked from the upper lateral branches of the vine, and kept covered out of the sun for a day or two to bleach them and thus improve the flavor.

The vine is native to Southeast Asia and is commonly grown around villages, railway stations in rural areas and in small home gardens. The vine needs a support.

An early authority (1894) recorded that in the manufacture of the *kris*, the Malay weapon, after arsenic was applied to bring out the pattern on the blade, betel leaves and lime were rubbed over the blade to remove the arsenic.

Plumeria

Plumeria spp.

Botanical family:
Apocynaceae

Charles Plumier first recorded this famous tropical flower in the 17th century. A misspelling occurred that has never been corrected. It originated from the more seasonal tropics of the New World, but has adapted itself to many other tropical countries.

The species found throughout the world are mainly *P. rubra* (in warm colors of red, gold and pink) and *P. obtusa* (with white or white and yellow flowers), although there are many unnamed and mixed color varieties. The plumeria can be a medium-sized tree or a dwarf shrub developed for pot culture. The branches are swollen and brittle and the whole plant produces copious amounts of white latex. The leaves are long and either pointed or blunt ended.

The bunches of flowers composed of individual flowers are fleshy and striking. They are very fragrant and are popular as temple offerings and cemetery plantings throughout Southeast Asia and India. In Burma and Singapore, the flowers used to be boiled, spiced, and eaten as a relish.

Hibiscus

Hibiscus spp.

The hibiscus in the form *H. rosa-sinensis* (bottom right) is the floral emblem of Malaysia, although it is not thought to have originated there, nor in China as its name indicates. It is a free-flowering, scarlet flower with a long, columned pistil and stamens projecting from its centre, set off by shiny green, saw-toothed leaves. It is not found in the wild and could be a very ancient hybrid.

Botanical family:
Malvaceae

The hibiscus, *H. schizopetalus*, below left, is from East Africa. It is distinguished by its recurved, slashed petals. It is a very old form and it may be a parent of the *H. rosa-sinensis*. Its dainty, pendant flower is very eye-catching.

These older forms are of more interest. In Old Malaya the petals of *H. rosa-sinensis* produced a black dye used for polishing shoes and as a cosmetic by Chinese women to darken their eyebrows. It also has some medicinal applications in Ayurvedic, Malay and Chinese medicine.

The disadvantage of hibiscus for decoration is that the flowers last for only a day, but they are plentiful and pretty.

Ironwood

Mesua ferrea

Botanical family:
Guttiferae

This is one of the sacred trees of India. It is found in a wide swathe across India, Sri Lanka, Burma, Thailand, Indochina, Malesia to Australia. It is a fairly common, although slow-growing, tree of the tropical rainforest and is a feature of many gardens and arboreta.

Ironwood is a tall, conical tree, up to 36 m in height, with glossy, very dark green leaves. It holds its lower branches as it grows and the sweep upwards of the conical form is therefore evident from quite close to the ground. The young leaves are also remarkable as they appear in flushes in bright red, folded together and pendant, and showing up quite clearly against the dark, mature leaves.

The *Mesua ferrea* has further glories. Its large, white flowers have four petals around a center filled with prominent yellow stamens and a strong sweet scent. The dried flowers are used to stuff pillows in India. In Burma, the yellow stamens are added to a traditional, white face powder to add a fragrance and golden glow to the powder.

Tree of Heaven

Amherstia nobilis

In many garden books written in English in colonial India before World War II, this tree was lauded as either the most beautiful or one of the most beautiful flowering trees in the world. It was discovered in Burma, and was named after the wife of the then Governor who was a keen botanist.

Botanical family:
Leguminosae

It is a medium-sized tree, between 5 and 15 m, but of a rounded, graceful habit. Its juvenile leaves are long, pendant and flushed an attractive pinky-purple-coppery color. The flowers are striking, each one about 7 cm, and held in a long, red-stemmed cluster of about 20 flowers. They are variously described as coral-colored or red-tipped with clearly distinct gold markings.

Amherstia prefers a tropical monsoon climate and displays its full glory only in its native Burma and, as older accounts have it, in Calcutta. As it is the single species of its genus, it rarely sets seed. Also, the seed, frequently sterile, has a short viability span.

49

Vanda Orchid

Vanda Miss Joachim

Botanical family:
Orchidaceae

This orchid was discovered in 1893 by Miss Agnes Joachim, an Armenian lady, in her garden in Pasir Panjang (Long Beach) in Singapore. Mr Henry Ridley, the then Director of the Singapore Botanic Gardens, determined the flower to be a natural hybrid as it was found between two other species of *Vanda*. Subsequently, it was named after its collector.

The flower is a pretty, open-faced bloom, pinky-mauve with violet and white. It is easy to grow, enjoys full sun and is free-flowering. Unfortunately, the blooms are not long-lasting, but their profusion makes up for this. After World War II, it began to lose its popularity as many other new orchid hybrids were being developed.

In 1981, of the many varieties of *Miss Joachim*, the form 'Agnes' was chosen to be the National Flower of Singapore. This rescued it from near oblivion, and commemorates very suitably a keen-eyed member of a once influential minority community in old Singapore.

Tiger Orchid

Grammatophyllum speciosum

This is the largest orchid in the world. It is native from Sumatra to Malaysia, the Philippines and Polynesia. *G. speciosum* is the largest of many species. It is an epiphyte found on the crowns of trees high above the ground often near streams. The stems when young are upright, but as they mature and elongate, they bend in graceful curves.

Botanical family:
Orchidaceae

The plant does not flower reliably, but when it does, it is a magnificent sight. There are many inflorescences some of which reach 3 m and can support 100 individual flowers each of which can be 15 cm in diameter. The sprays are held upright and the flowers are clearly seen. They are ochre or greenish yellow with markings variously described as dull orange-brown to maroon-purple.

In cultivation the plant is not very demanding. It must have full sun and will grow in a raised, free-draining bed made of brick fragments and charcoal mixed with decaying leaves. Its stiff, upward-growing feeding roots catch the detritus of organic matter on which the plant feeds.

Amazon Waterlily

Victoria amazonica

Botanical family:
Nymphaeaceae

This, the largest of all waterlilies, was discovered in the early 19th century in the backwaters of the Amazon river. An Englishman, John Lindley, some decades later described it in the detailed scientific way required for botanical classification and patriotically named it *Victoria regia* in honor of Queen Victoria. Today, it has been renamed *V. amazonica*, and it reigns supreme in large ponds and lakes in Botanic Gardens around the world.

The gigantic leaves reach a diameter of 2 m. They are round floating trays with an upturned rim of 7.5 cm. The leaves are held together by prominent veins on the prickly underside; these veins circulate food and water and hold air, thus giving the huge leaves buoyancy and strength.

The football-sized flowers open white, and over two days, change to pink then maroon. The pollinator is a scarab beetle, attracted to the warmth produced by the breakdown of starch particles in the floral cells. The pollinated flower sinks into the water where the seeds develop.

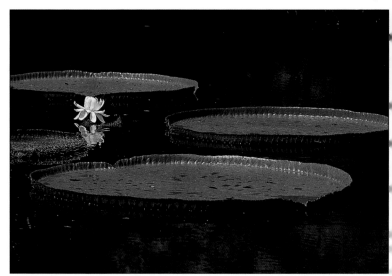

Sacred Lotus

Nelumbo nucifera

The sacred lotus, a handsome aquatic plant and one of the most revered flowers in the world, is extensively cultivated in Asia both commercially and in countless home water jars. It is the dominant flower of the Buddhist Way and is significant in the worship of the Hindu Brahma. In Taoism, it is the emblem of one of the Eight Immortals. It symbolizes purity because the flower stem rises high above the mud in which it grows. The buds look like hands put together in prayer.

Botanical family:
Nelumbonaceae

It is grown from a rhizome and from seed. The shallow, bowl-shaped leaves are rounded, between 30 to 60 cm in diameter, and are covered with a waxy bloom that reflects light from drops of water like quicksilver. The large, scented, pink or white single flowers have many overlapping petals forming a deep bowl. They are bisexual and easily pollinated.

Fruits are inverted cones with individual seeds embedded in the top flat plane. All parts of the lotus can be eaten.

Keng Hwa; Lady of the Night

Epiphyllum oxypetallum

Botanical family:
Cactaceae

In Malaysia and Singapore this member of the cactus family can sometimes be seen decorated with small red ribbon bows. This indicates that the plant has given somebody 'lucky numbers' for the many multi-digit gaming systems in these countries.

Native to South America, it arrived in the area in 1921 via Canton in China, thus acquiring a Chinese name. It has the flat bladed, notched, modified cactus leaves as stems and a rather untidy habit. The flower buds develop up to a certain stage and then await the impetus of a cool, wet day which stimulates them to flower 24 to 25 days afterwards.

In a healthy plant many flowers are produced simultaneously and gregariously, ie all the plants in the same district will bloom at the same time. They bloom at night, opening at about 11 o'clock, and fade with the dawn. The large, scented flowers are creamy to creamy-pink with prominent stamens, and are held on the stem in a curious way.

Keng Hwa is propogated from cuttings.

Cannonball Tree

Couroupita guianensis

This native of South America is now found all over the tropics as a specimen tree as well as a curiosity. It grows to a height of 35 m with a straight trunk and produces its flowers and fruits from its lower trunk at human height.

The whole inflorescence emerges from the trunk and can be 3 m in length. The 6 cm flowers attached to this stem are creamy-pink, sometimes a deeper red, with prominent stamens. The fruits that follow are also large, round and hard, and look like brown cannonballs. They are usefully described as being smaller than a man's head. Another explanation for the name is that the fruits when knocked together make a sound like cannon fire.

The tree is interesting and attractive in a rather bizarre way. It flowers and fruits simultaneously. Some authorities state that the flowers produce a strange scent. When the main flowering is over, the tree is then hung about with the brown fruits. They take about a year to ripen and the seeds are embedded in an unpleasant smelling pulp.

Botanical family:
Lecythidaceae

Fig. 81.—Rafflesia Arnoldii, reduced from photograph of living flower.

Botanical family:
Rafflesiaceae

Rafflesia

Rafflesia spp.

If there were flower Olympics, the *Rafflesia* would win every time, as it is the largest flower in the world. To date, 16 species have been recorded, almost all in the Malesian region, with the main points of sightings in Sabah, East Malaysia, the Malay Peninsula and in Sumatra. However, three Borneo species are suspected to be extinct, as no reports of them have been published for 90 years.

The plant—and particularly the flower bud which is the size of a small cabbage—was known to aboriginal and Malay medicine well before the 19th century, when it was discovered for the European scientific world by the indefatigible Stamford Raffles. Raffles was in Sumatra in 1818 with his friend the naturalist, Dr Arnold, who actually found it in the area of a town called Manna. It was later named *R. arnoldii* (see top left). It is the largest of all the *Rafflesia* flowers, with a diameter of 80 cm and a weight of 7 kg.

Rafflesia spp. are remarkable, not only because of their size, but also because the flower is the only part of the plant that is visible. It does not have any of the parts that are considered 'normal' for a plant—for example, leaves, stems and roots. Rather, it is a parasite, its host being a vine belonging to the *Tetrastigma* genus. As with many features of the species, it is not known how the vine is parasitized, but it is postulated that the seed enters its host by means of an injury. The buds take over a year to mature and aborted buds are common.

The structure of *Rafflesia* flowers is common to all species. It can be described somewhat unscientifically as a large basin with a rim curving over the center, surrounded by five fleshy lobes, the equivalent of petals. Within the basin, properly called a diaphragm, is a fleshy disc on a thick column containing the sexual parts of the flower (see opposite bottom). The pollinators are thought to be carrion flies. The pollinated flower sets seed, and the agents which distribute the seed possibly include wild pig, shrews, squirrels, termites, and ants.

57

Botanical family:
Nepenthaceae

Pitcher Plants

Nepenthes spp.

Linnaeus named this plant after a Homeric mention of a drug for sleep and oblivion. This however cannot in reality describe what happens to insects and small creatures that are unfortunate enough to be lured into the trap that is the pitcher plant. They slide down the slippery inner slope of the pitcher and are prevented from climbing up again by the waxy walls covered in scales. Death is a slow and—one fears—painful process. This is because ants and other small insectivorous creatures (as well as rats and toads that are tempted there by the fluid in the pitchers during times of low rainfall) are consumed by the digestive juice in the pitcher. This juice is hardly diluted by rainwater as the pitcher produces a lid to cover the liquid. One conjectures that sleep and oblivion would be merciful for creatures trying to escape from the trap.

Paradoxically, there are other creatures, such as mosquito larvae, that have a symbiotic relationship with the plant. They live in the fluid and partake of the meal being digested by the plant, but are not digested themselves. It is presumed that these larvae form the bait by which other insects are lured into the pitcher. There are also reports of a spider which spins its web at the mouth of the pitcher, and, when threatened, flees into the fluid and emerges later, unharmed.

Nepenthes are undershrubs or climbers. Both the pitcher and the lid that protects it are modified leaves. Their flowers are nondescript, but their pitchers which come in various sizes are extremely flamboyant. All *Nepenthes* are green with some red in them. Some are wonderfully speckled and blotched. Some pitchers are globular, some elegantly stretched out, and there are tremendous variations in the lip of the pitcher.

The distribution of *Nepenthes* is quite widespread, but the most number of species is in Borneo. Many of the species are protected under international rules, but there are cultivated varieties.

Dillenia

Dillenia spp.

Botanical family:
Dilleniaceae

Dillenia suffruticosa (below) and *Dillenia indica* (on left) are a large shrub and a small tree respectively.

They have large, rough leaves with toothed edges, a distinct central vein, and, in many species, beautifully marked parallel secondary veins. The leaves appear to be quilted in neat long rows. Fresh leaves were traditionally used in markets as a wrapping for fresh beancurd and for cooked food that did not require liquid.

Dillenia flowers are white or yellow. They are large with prominent stamens. In *D. suffruticosa* the fruit is a capsule that opens to reveal red seeds in a red aril, attractive to birds. *D. indica*, called the elephant apple, has a large, edible watery fruit with an acid flavor.

D. suffruticosa is a plant that lives by an exact time-table. Its flowers open and close at precisely timed intervals daily. The fruits set on the 36th day after the petals fall. Without fail, the fruit and flowers open at 3am in readiness for the dawn. The shrub can live for up to 100 years.

Java Olive

Sterculia foetida

This is a tall tree with a straight round bole that reaches 40 m at maturity. The crown is spreading and the leaves are palm-shaped holding many leaflets. The tree sheds its leaves periodically. It is a minor timber tree useful for building packing cases and temporary shelters.

Botanical family:
Sterculiaceae

The flowers are small and inconspicuous but make their presence known by their unpleasant and pervasive odor. It is this characteristic that gives the tree its specific name *foetida*. The attractive fruits appear in large follicles (pods) about 2–5 cm broad, with a tough, red skin in clusters, like an outspread hand. The ripe pods split open revealing two to three roughly round black seeds. These, either roasted or raw, are edible in moderation.

In Bali, *Sterculia foetida* is frequently seen around temples dedicated to Durga, a deity in the Hindu pantheon who alleviates suffering and sorrow and releases humans from material desires and human attachment. Previously, the red skin of the seed pod was used in batik dyeing.

Ficus

Ficus spp.

Botanical family:
Moraceae

Ficus are a very large family of plants of the tropics numbering about 800 species. They are roughly divided into the stem *Ficus* and the strangling *Ficus*. All *Ficus* produce fruit in the form of a fig, the stem ones from the trunk and branches and the stranglers from the base of the leaves.

The plant shown on opposite top is *Ficus benjamina* or the *waringin* tree of Indonesia. It is a strangler. It starts off in life as a seed, usually scraped off the beak of a bird on to a tree. For a while it is an epiphyte, then it begins the slow destruction of its unwitting host. It puts out roots which grow rapidly towards the ground where they quickly establish themselves. All the while, the plant tightens its grip around the host tree. Eventually the host dies, and the fig, now growing with its own roots anchored firmly in the ground, is seen as a thick mesh around the remains of the original tree. The *Ficus* now proceeds to grow by sending down more aerial roots that will continue to prop up its increasing size.

In Indonesia the *waringin* is said to be the abode of spirits. It is almost never cut down, but if necessity demands that it has to be felled, a number of placatory ceremonies are conducted before the ax falls. Abstracted renditions of this tree, unidentifiable as to species, are used as logos for all manner of commercial ventures.

Another famous strangler is the *Ficus benghalensis*, under whose shade the army of Alexander is said to have rested during the invasion of India. This is the real banyan tree, a term now loosely applied to the *F. benjamina*. It takes its name from the *banians* or itinerant peddlers who rested under its shade.

The stem *Ficus* (opposite below) is the revered *Ficus religiosa*. It was while seated under this tree that Bhudda attained enlightenment. Devotees will never cut it down. It frequently starts life growing out of a crack in a wall. The tree is single-stemmed, has heart-shaped leaves with a long distinctive drip tip, and can become very large.

Index